What's the weather like today, like today, like today?

What's the weather like today?
Today is cloudy.

What's the weather like today, like today, like today?

What's the weather like today, like today, like today?

What's the weather like today?
Today is windy.

What's the weather like today, like today, like today?

What's the weather like today?
Today is snowy.

What's the weather like today, like today, like today?

What's the weather like today?
Today is sunny.

What's the weather like today, like today, like today?

What's the weather like today?
Today is foggy.

What's the weather like today, like today, like today?